PO
PETS

Written by
William O'Byrne

HORWITZ
MARTIN

Contents

Chapter 1

People And Pets

Pets are animals that make good friends. They may be useful or they may be interesting. They are fun to have around.

The first popular pets people ever had were dogs. There are very old paintings from the Stone Age of people and dogs.

Nowadays, all kinds of animals are kept as pets. Popular pets are dogs, cats, fish and birds. Many people have lizards, hamsters, pigs, goats, spiders, mice: almost anything!

It doesn't matter what the animal is, as long as you look after it properly and have fun with it.

Chapter 2

Dogs

All dogs are related to wolves. They may look different to wolves, but dogs often behave the same way as wolves do.

Dogs are very social animals. They like to be with other dogs or people.

All dogs need plenty of exercise to keep healthy. Puppies need more care, such as training and good food.

If you like big dogs, you can get a Saint Bernard. It can weigh up to 100 kilograms! If you want a little dog, the tiny Chihuahua can weigh less than one kilogram.

Dogs are used in many ways, such as protecting and rescuing people.

Did you know ...?

Before people had heaters, small dogs might sit on their laps to keep them warm. These dogs are called lap dogs.

Chapter 3

Horses

Horses are another popular pet. But, for thousands of years, horses have mostly been bred to help people with different jobs.

The knights of the Middle Ages needed big, strong horses to ride on, because their armour was so heavy. Other horses were used for farm work, pulling carts and ploughs.

Some horses were bred to be fast runners. A really fast horse can gallop at 70 kilometres an hour. Today, most horses are used just for riding and racing.

Chapter 4

Cats

Cats are the most common pets kept today. There are more than 500 million domestic cats kept by people around the world. They can come in all shapes, sizes and colours.

Many people like to have cats as pets because they are friendly and furry.

Cats are not like dogs, because they don't mind being alone.

Cats are nocturnal. This means they like to rest during the day and go out at night to explore and hunt.

They are carnivores, which means they only need to eat meat.

Domestic cats are related to lions, tigers and cheetahs.

Cats were very popular with the early Egyptians. They made many statues of their favourite pets. The early Egyptians also kept monkeys, lions, dogs and geese as pets.

Chapter 5

Birds

If you were a pirate, what would your perfect pet be? Perhaps a talking parrot that could sit on your shoulder?

Many people keep budgerigars (or budgies), cockatiels and parrots because they can talk! These birds may not know what they are saying, but they can repeat things like nursery rhymes, their names and anything they else they may hear!

So don't say anything rude in front of a parrot, or it might repeat it!

P

Parrots are clever and friendly. However, they can bite if they get annoyed. Parrots like to eat fruit, nuts and vegetables.

Parrots can live for a long time—maybe up to 100 years. Some cockatoos can live for over 50 years.

Budgies can live to 12 years old and cockatiels up to 15 years.

100 Years!

If you are not a pirate with a parrot on your shoulder, your birds need a cage. Many birds that live in cages can be trained to come out and fly around the house and go back to their cage to sleep.

Birds are popular pets. Not only are they clever and colourful—they can also flutter and fly!

Chapter **6**

Fish

Fish are good pets for people who don't have a lot of room.

Goldfish are cold-water fish that can live in ponds or aquariums. There are many types of goldfish.

Some have unusual names like Fantail, Lionhead, Bubble Eye and Shubunkin. They can also look unusual too!

Tropical fish need warm water to live in. They are usually very colourful. One type of tropical fish, the Oscar, will swim over to its owner. It will ignore people it does not know.

Some fish lay eggs. Others, like guppies, give birth to live young fish (or fry). A guppy can have up to 50 fry every four to six weeks.

Looking after fish means being careful with their water. The water in the aquarium must be the right kind of water, at the right temperature. Many fish die because people use water straight from the tap.

Goldfish don't like water hotter than 23 degrees. Tropical fish need water that is between 24 and 27 degrees. Don't put goldfish and tropical fish in the same aquarium!

An aquarium with brightly coloured tropical fish is lovely to look at!

Some fish, such as male Siamese Fighting Fish, like to fight and eat other fish. They are lovely to look at but, if you put two of them in the same aquarium, they will fight until one of them dies.

Before you put fish in a tank, find out from the pet shop owner which fish get along together.

You will have to feed your pet fish every day.

You will also need to clean the tank and change the water regularly to keep your fish healthy.

Chapter 7

Strange Pets

Some people keep giant snakes, such as boa constrictors or pythons as pets.

Others like big spiders, such as tarantulas!

Reptiles like goannas, lizards and tortoises are also popular as pets, because they are interesting creatures.

Whatever you choose to have for a pet, make sure you can give it the care it needs.

Pets live for a long time and they need you to look after them. When you go away on holiday, remember to arrange proper care for them.

In return, they will give you lots of enjoyment, and be a friend who is *always* there for you.

Index

Glossary

aquarium
A fish bowl or fish tank.

bred
Made better, by carefully choosing an animal's parents.

domestic
Living with people.

Middle Ages
A time between 600 and 1,500 years ago.

ploughs
Farm tools used to turn over soil.

popular
Something a lot of people enjoy.

related
From the same family.

social
Liking to be part of a group.

Stone Age
A time more than 10,000 years ago.